This edition is the fourth and last in the series 'Hendon as it was', and is devoted entirely to Hendon's long association with flying.
No attempt to compile a pictorial record of Hendon during the first quarter of this century would be complete without special reference to aviation.
This little book does not pretend to have the erudition of a history, nor is it a complete pictorial record. It is an attempt—within modest limits—to record pictorially some of the events, the personalities and machines that made Hendon an historic centre for aviation.
I have concentrated mainly on the years 1910-1930 and endeavoured to provide a glimpse of Hendon Aerodrome during this period.

I should like to express my gratitude to a number of friends for their help and encouragement during the past year and to the many people who have contributed to my knowledge of Hendon.
Thanks are due to the Borough Librarian, Mr. S. J. Butcher, F.L.A., F.R.S.A., and the London Borough of Barnet Library Services for allowing me access to material in the local history archives: Mr. J. M. Bruce, M.A., F.R.Hist.S., A.R.A.E.S., Keeper of Aircraft and Aviation Records, Royal Air Force Museum, Hendon. The Hendon Times and Daily Mirror. My special thanks are due to Mr. E. Nicholls, Mr. John Chaplin, Mr. Brian Girling and Mrs. Gladys Carr.
Much of the material used is from my own collection; however, I am indebted to Syndication International for allowing me to reproduce a number of photographs in this book; and to the late Squadron Leader Herbert Allen, who shortly before his death, was kind enough to talk to me at great length about the early days of flying at Hendon.

C.R.S.

"A week-long search for the future home of his flying school brought him one morning to Hendon Hill, and from near the church he gazed down on several hundred acres of pasture bounded on one side by the embankment of the Midland Railway. This was his Darien, and he knew he need look no farther. Here only a few miles from the heart of London, he would one day establish a great centre for British aviation." — January, 1910.

From Claude Grahame-White by Graham Wallace.

Photo. Above — Hall Lane now Greyhound Hill overlooking Hendon Aerodrome. — c. 1910
Photo. Right — A Morane-Saulnier monoplane flying low over rural Hendon. — c. 1912

Flying really began at Hendon in 1910, less than seven years after the Wright Brothers made the first powered flight in December, 1903.
In 1910, the Daily Mail offered a prize of £10,000 for a flight from London to Manchester. In April of that year the French flier Louis Paulhan took off from Hendon and won the prize. His rival in the race, an Englishman, Claude Grahame-White flew from Wormwood Scrubs and although Paulhan won the race the honours were divided equally, for in an attempt to regain lost time, Grahame-White took off an hour before dawn on April 28th, 1910 and so became one of the first men to fly at night. Both Grahame-White and Paulhan flew Farman biplanes, with their multitude of struts and wire bracing; these were aptly nick-named box-kites. They were fragile machines, the pilot sat in an unprotected bucket seat, with his feet on the rudder bar. The engine at his back was a 50 h.p. Gnome driving a "pusher" propeller.
A few months after the London to Manchester race Claude Grahame-White acquired 207 acres of pasture land at Hendon to start what he called 'The London Aerodrome'. He persuaded M. Louis Bleriot (of Channel crossing fame) to join him in his aviation activities at Hendon and start a flying school.

Left — The London Aerodrome Hendon, 1910 (showing the shed used by Louis Paulhan)
Right — Claude Grahame-White and below. Louis Paulhan, 1910.

"M. Louis Paulhan and Claude Grahame-White—Hendon is interested in these two airmen, for both have visited the aerodrome at Colindale. During the latter part of last year, when aeroplanes were not so common as they are now, Mr. Grahame-White tried a new machine at Colindale; whilst the flight of M. Paulhan from Hendon to Manchester is still fresh in the memory of Hendon residents" . . .

Finchley & Hendon Times — August 19th, 1910

"A good deal of work has been done during the week at the new aerodrome at Colindale, Hendon and within the next few days we may witness some flights from a place which is being much talked of among aviators. The Bleriot aviation school has been established at Colindale."

Finchley & Hendon Times — 30th Sept., 1910

"On the grounds of the London Aerodrome Co. at Colindale, Hendon an informal inauguration took place on Saturday of the aviation school which M. Louis Bleriot proposes to found in this country. The site provides for a course of from 2—3 miles. At present there is shed accommodation for 8 machines, one of the sheds being that from which M. Paulhan started on his London to Manchester flight. Three of the sheds have been taken by an aeronautical syndicate which is also founding a school, two by M. Bleriot and the others are available for the purposes of the Aerodrome Co."

Finchley & Hendon Times — 7th Oct., 1910

'AEROPLANE OVER HENDON'

"On Wednesday afternoon Hendon was thrown into a state of excitement by the sight of an aeroplane passing right over the town, a sight which no doubt will be quite common now that the aviation school is in full swing. The adventurous airman was M. Pierre Prier, who has come especially to Hendon to give instruction in Mons. Bleriot's school, and so far he has proved himself to be a flier of no mean order. On several occasions he has given some really excellent exhibitions but the most wonderful so far as he has flown at Hendon was that on Wednesday. Seeming to come up from Colindale, he passed over the Burroughs and turned again, running parallel to Parson Street. He then made towards Colindale again. Altogether M. Prier was in the air for about three quarters of an hour. The machine used was a monoplane of the Bleriot design. The school at Colindale is fast gaining a name in the Metropolis and we understand that some of the pupils have given excellent results."

Finchley & Hendon Times — Oct., 1910

In 1911 Grahame-White, assisted by his friend Richard Gates, began to develop Hendon into an aviation centre. Land was levelled, and trees felled. When completed the airfield was oval-shaped, two miles in circumference, and included an office block, various workshops and 17 hangars. The Grahame-White School of Aviation was established and Clement Greswell appointed chief pilot assisted by Charles Hubert and J. V. Martin.

A man of vision, Grahame-White was quick to perceive the aeroplane's military potential. In May 1911, a demonstration was given at Hendon to the Parliamentary Aerial Defence Committee. Members of Parliament—including Mr. Winston Churchill—were shown the "scouting, bomb dropping and dispatch carrying" abilities of the aeroplane.

Grahame-White also began to stage air races and displays at Hendon. From 1911 thousands came to see such heroes of the new age of aeronautics as—A. V. Roe, S. F. Cody, T. O. M. Sopwith, Henri and Maurice Farman, Gustav Hamel, B. C. Hucks and Claude Grahame-White himself—and to witness events that were milestones in the progress of aviation.

To the surprise of many in the financial world, the gate receipts for the first year totalled £11,000.

Photo. Top Right — Louis Bleriot (1872 - 1936)

Residents of Hendon were privileged to be part of the beginning of flying in this country and Sunny Hill Fields overlooking the aerodrome provided a perfect free grandstand view.

Thousands walked by lane and footpath—for in those early days Hendon lay well beyond the boundaries of Greater London. The occasions were often competitive, a race for a cash prize or trophy. In the three years preceding the First World War, Hendon became a popular rendezvous at the weekend for many thousands of Londoners. The pilots were idolized by the crowds who came to watch and they were acclaimed rather like Film and TV stars are today.

Colindale Avenue, then muddy and potholed, was the only approach road to the Aerodrome and on race days was thronged with people on their way to the airfield. Some of the residents sold tea and refreshments to the visitors and street vendors would do a brisk trade selling souvenirs, programmes, leaden toys of aeroplanes and picture postcards of the aviation personalities of the day.

Left — Picture Postcards of the Hendon aviators and above — Farman-biplane flying over spectators at Hendon. — c. 1911

Left — Crowds at Hendon.
— c. 1911
Right — Claude Grahame-White and (below) signing autographs for some of his many admirers.
— c. 1911

Claude Grahame-White (1879 - 1959)

A Bleriot trainer plane at Hendon; this machine did not fly.
(note the shortened wings). — c. 1913

By 1911 there were a number of flying schools at Hendon with instructors, mechanics and pupils of many nationalities, British, American, French, Belgian, Italian, Swiss and Russian—an international community—drawn together by a new language—aeronautics. Instruction in those early days was very meagre, the instructor would put the pupil into an aircraft, probably a 50 h.p. Farman-biplane or 'box-kite' and explain the simple controls and the way to handle the machine. There were no dual controls and so when the pupil—after many hours of taxi-ing to and fro around the airfield—made his first solo flight, the instructor would have to use hand signals or shout.

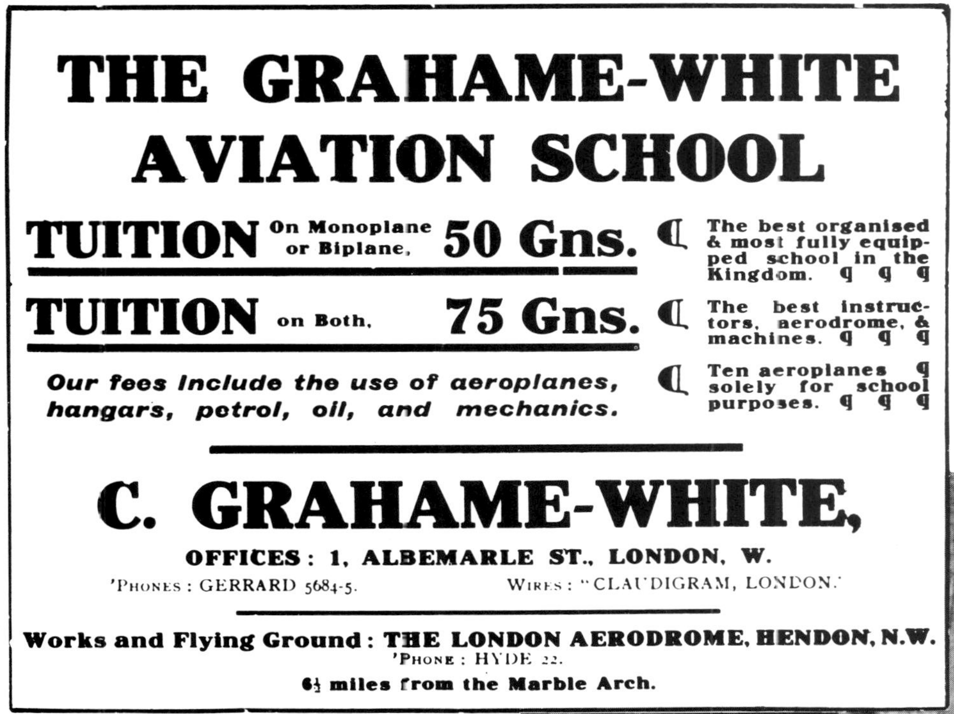

An advertisement for Grahame-White's Flying School. — c. 1911

Right — Part of Grahame-White's Flying School at Hendon. Photograph shows an aircraft being repaired in one of the workshops. — c. 1911

GUSTAV HAMEL

Apart from Grahame-White, the most popular airman at this time was probably Hamel. Handsome, young and extremely skilful, he drew large crowds wherever he appeared. Visitors to Hendon were always assured of a magnificent performance if Hamel was flying.

In 1910, when he was 21, Hamel went to France and learned to fly there. Less than a month after gaining his pilot's certificate by flying a 50 h.p. Morane monoplane, he was taking part in air races. In 1911, he won a race from Brooklands to Brighton and in September of the same year Hamel became the first aerial postman when he flew Europe's first official air-mail from Hendon to Windsor. The weather was appalling when he climbed into the open cockpit of his little Bleriot and the mailbags containing special postcards and letters bearing the postmark 'First United Kingdom Aerial Post' were placed on board. When he took off he had to battle with a wind of 56 miles per hour.

The Daily Mirror reported the flight on 12th September—HAMEL'S FLIGHT.

> *"Mr. Hamel's fine flight from Hendon to Windsor in inaugurating the aerial post service on Saturday seems to have been an even more magnificent feat than it appeared. According to a statement issued by the organisers from the Aerial Post Office, Hendon Aerodrome, the official time occupied by Mr. Gustav Hamel in his flight from Hendon to Windsor was —*
>
> TEN MINUTES
>
> *Notwithstanding unfavourable winds, Hendon was left at* 4.58 *p.m. and Windsor reached at* 5.08 *p.m.*

The following year, 1912, Hamel flew with a passenger, Miss Trehawke Davies from London (HENDON) to Paris. He left Hendon at 9.38 a.m. in a two-seater Bleriot monoplane and arrived in Paris at 5.55 p.m.

On 20th September 1913, Hamel won the 2nd United Kingdom Aerial Derby, starting and finishing at Hendon. His time for the 94.5 mile course was 1 hour 15 minutes 49 seconds. This brilliant pilot died while piloting a new racing Morane monoplane from France to England in May 1914.

Air races were a regular feature at Hendon and those who took part in them became popular figures. They would fly their machines—usually Farman and Grahame-White biplanes—round and round the aerodrome banking steeply at the pylons and always flying low. Among the famous Hendon pilots were Gustav Hamel, Lieut. J. C. Porte, W. L. Brock, L. W. F. Turner, Louis Noel, Marcus D. Manton, Sydney Pickles, Pierre Verrier, R. H. Carr, W. Birchenough, J. Lawrence Hall, L. A. Strange, Bentfield C. Hucks and Claude Grahame-White.
Bentfield Hucks was the first English pilot to loop the loop. Hucks had taught himself to fly in 1910 and in 1913 at Hendon in a Bleriot monoplane he looped the loop.
By 1914 demonstrations of looping had become a regular feature at Hendon (see page 2). Hamel, Hucks and Manton were the foremost exponents in the art of aerobatics and during the year preceding the first World War visitors to Hendon were treated to many daring displays of trick flying. On one occasion it is said Gustav Hamel swooped very low over the crowds and the tail of his plane knocked a policeman's hat off.
Grahame-White organised an "upside down" dinner at the Royal Automobile Club. The guests of honour were Hamel and Hucks. The guests ate their way backwards through ten courses beginning with cigars and coffee.

"DAILY MAIL" AEROPLANE CIRCUIT. 1912
MR B. C. HUCKS ON HIS 70 H.P. BLERIOT.

EVENT BOARDS. The progress and results of events on the daily programme will be announced by megaphone in each enclosure.

BOMBS. The commencement and finish of the daily programme will be signalled by the firing of a bomb.

FLAGS. The aviators will be started by the dropping of a white flag.

The completion of the course by each competitor will be signalled by the waving of a red flag at the starting line.

The fouling of a pylon by a competitor will be signalled by the waving of a black and white flag by the pylon keepers of the two succeeding pylons. and the competitor must return and pass the fouled pylon in a proper manner, otherwise he will be disqualified.

SIGNALS. Prior to the start of each race a black inverted cone will be hoisted at Pylon No. 1, opposite the judge's stand. After a race is started, a black ball will be hoisted in addition to the cone, indicating that a race is in progress. At the finish of each race the cone will be hauled down, leaving the ball to indicate that the results are being announced.

BELLS. The start of each event and the last lap of each race will be announced by the ringing a bell in each Enclosure; on the judge's box in the Paddock Enclosure, on the Grand Stand in the 1/- Enclosure, and at the front of the Refreshment Pavilion in the 6d. Enclosure.

All other announcements will be made at each enclosure by megaphone

A Nieuport monoplane circling one of the famous Hendon pylons. The Grahame-White hangars can be seen in the background. — c. 1912

Aviators at Hendon.

Name.	Nationality	Machine. Make.	Type.	Engine.
PIERRE VERRIER ...	F	H. Farman	Biplane ...	80 h.p. Gnome
		Maurice-Farman	Biplane ...	70 h.p. Renault
LOUIS NOEL	F	Grahame-White	5-Seater Biplane ('Aero-Bus')	100-h.p. Green
R. H. CARR	E	Grahame-White	Tractor-Biplane ...	50 h.p. Gnome
J. LAWRENCE HALL	E	Avro ...	Biplane ...	50 h.p. Gnome
F. W. GOODDEN ...	E	Caudron ...	Tractor-Biplane	60 h.p. Anzani
E. BAUMANN	S	Wright	Biplane ...	40 h.p. Wright
RICHARD T. GATES...	E	Grahame-White	Biplane ...	50 h.p. Gnome
L. A. STRANGE ...	E	Bleriot	Monoplane	80 h.p. Gnome
		Grahame-White	Biplane ...	50 h.p. Gnome
W. BIRCHENOUGH...	E	Grahame-White	Biplane ...	50 h.p. Gnome
A. E. BARRS	E	Grahame-White	Biplane ...	50 h.p. Gnome
R. J. LILLYWHITE ...	E	Grahame-White	Biplane ...	50 h.p. Gnome
N. HOWARTH ...	E	Grahame-White	Biplane ...	50 h.p. Gnome
E. F. NORRIS	E	Grahame-White	Biplane ...	50 h.p. Gnome
AND C. GRAHAME-WHITE	E	Maurice-Farman	Biplane ...	70 h.p. Renault

Left — Claude Grahame-White with a lady passenger in a Farman biplane. — c. 1911

Below — Crowds at Hendon watching competitors in the Daily Mail air race 1911.

Right — S. F. Cody (1861 - 1913)

Samuel Franklin Cody, a popular figure at Hendon during those early years. He is seen here at the controls of his own plane, seated behind is Mrs. Cody.

Cody's plane became a familiar sight at Hendon. It was a huge contraption of spars and flaps and was nicknamed the flying 'cathedral'. When it touched down, the clatter could be heard half-a-mile away.

Cody died in 1913 when he crashed while preparing to enter the Northcliffe competition for a seaplane flight around Britain. Photo. — c. 1911

NOTE. Cody is credited with the first sustained powered flight made in Great Britain—1,390 feet, 16 October, 1908.

Born in America he took British citizenship in October, 1909.

Below — Claude Grahame-White climbing down from his plane after crash-landing at Hendon. 1911.

Above — Pierre Verrier, chief pilot for the Farman Company in a Farman Longhorn biplane at Hendon. 1912.
Inset — James Valentine's Bristol monoplane at Hendon. — c. 1912

Left — S. Cody's famous biplane seen regularly at Hendon during 1912-13.

Below — A speed contest at Hendon in 1912. Left: Gustav Hamel, centre: T. O. M. Sopwith and right: S. F. Cody.

THE AERIAL DERBY

In 1912, the first great Aerial Derby started and finished at Hendon. The London Daily Mail sponsored the competition with a 100 guinea trophy and cash prizes of £460 (First Prize £250). The course was a circuit of London and attracted the world's leading aviators.
It was won by T. O. M. Sopwith in a 70 h.p. Bleriot monoplane, in a time of 1 hour 23 mins. 8.4 seconds. Forty-five thousand paid to see the race at Hendon and it was estimated a further half-a-million watched the competitors on their way round the 81 mile course.

Left — Huge crowds on their way to see the start of the first Aerial Derby at Hendon, 1912.

Below — T. O. M. Sopwith on his Bleriot after winning the first Aerial Derby at Hendon, 1912.

THE FIRST AERIAL DERBY.

THE AERIAL DERBY GOLD CUP—1914.

With interest in aviation growing all the time, the months preceding the beginning of the First World War were extremely busy at Hendon. Thousands came to see the many events, notably the air races, Hendon-Manchester-Hendon, Hendon-Paris-Hendon and the third Aerial Derby. The Grahame-White Company was expanding rapidly. In addition to the flying school an increasing number of aircraft were being assembled in the Company's factory. Apart from Grahame-White designs, there were Breguets, Morane-Saulnier and Nieuports being produced under licence.

Inset Right — W. Brock seen here in a Deperdussin monoplane. Brock won the 3rd Aerial Derby in a Morane-Saulnier, with a time of 1 hour 18 mins. 54 secs.

"First Europe, and then the globe, will be linked by flight, and nations so knit together that they will grow to be next-door neighbours. This conquest of the air will prove ultimately, to be man's greatest and most glorious triumph. What railways have done for nations, airways will do for the world."
CLAUDE GRAHAME-WHITE and HARRY HARPER (1914)

"Mr. Claude Grahame-White as the courageous champion of an act as yet in its infancy, has set his mark on Hendon. The name of the town is flashed over the face of the earth and water to the confines of civilization by the exhibits which take place over the aviation grounds."
Hendon & Finchley Times — 16th January, 1914

Above — Spectators watching an air race at Hendon, a Breguet biplane can be seen crossing the starting line. — c. 1914

PROBABLY motorists are not more interested in flying than anyone else, but nevertheless, a remarkable feature of the Easter Meeting at the Hendon Aerodrome, was the number of people who motored down and took their cars into the enclosure. One explanation is that there is no more comfortable vantage point from which to see the aerial evolutions than a car. To lean back among the cushions and watch Hamel, for instance, climbing to a height of 4,000 feet, and then making twenty-two consecutive "loops" in his descent, affords a series of thrills which can certainly be enjoyed nowhere else in such comfort.

The arrangements for cars in the enclosure at Hendon are really excellent, and there is no possibility of getting wedged in among the crowd, as so often happens at other places where special arrangements for cars are supposed to be made. There is plenty of space, too, and it is the easiest thing possible to get out at any moment when duty or pleasure necessitates return to town. Moreover, an excellent tea is brought up to the car at a very moderate charge. Certainly, for those who want to motor somewhere at a time when the country roads are crowded with holiday-makers, there can be no better use to put a car to for an afternoon than to take it into the enclosure at Hendon.—*The Court Journal, April 16th, 1914.*

Top Right — Marcus D. Manton and Lieut. Stellingwerf on a Grahame-White biplane used for Lewis Gun tests in 1913.
Marcus Manton described in picture postcards as "the wonderful boy aviator" was an instructor and test pilot for the Grahame-White Company. In 1914 he teamed up with B. C. Hucks in aerobatic displays at Hendon.
Manton qualified as a pilot at the Grahame-White School in 1912 at the age of 18.

THE "Aerobus" holds four passengers comfortably! Why not take your party for a flight?

Photo: "Daily Mirror."
The Grahame-White All-British Five-Seater Biplane ("The Aero-Bus.")
Passenger Flights may be booked from £2 . 2 . 0 each Passenger.

TO BOOK A PASSENGER FLIGHT!

PASSENGER FLIGHT CHARGES.

(a).	PASSENGER FLIGHT. Two circuits of the Aerodrome	£2 . 2 . 0
(b).	EXTRA FLIGHT. Two higher and wider circuits of Aerodrome	£3 . 3 . 0
(c).	SPECIAL FLIGHT. Outside the Aerodrome, in the direction of Edgware, returning towards the Welsh Harp	£5 . 5 . 0
(d).	DOUBLE FLIGHT. Four circuits of the Aerodrome	£4 . 4 . 0
(e).	MONOPLANE PASSENGER FLIGHTS from	£3 . 3 . 0

Cross-Country Flights.

ELSTREE and BACK—about 16 miles	£10 . 10 . 0
ST. ALBANS and BACK—about 27 miles	£20 . 0 . 0
BROOKLANDS and BACK—about 38 miles	£26 . 5 . 0

Other Cross Country Flights.

Per mile, if return journey is made	15/-
Per mile, for outward journeys only	20/-

All Flights are given subject to the conditions issued with Flight Tickets, and weather and all other circumstances permitting.

Flights can be booked at the Offices under the Judge's Stand, or at the Company's Offices, The London Aerodrome, Hendon, N.W. Telephone 120 Kingsbury (4 lines), and at 166, Piccadilly, W. Telephone 5955 Mayfair

Below Right — R. H. Carr standing alongside a Grahame-White 'Lizzie'. This machine was designed and built mainly for exhibition and aerobatic flying. In March, 1914, at Hendon, Carr looped in the 'Lizzie'.,
NOTE — This plane's fuselage was that of a Morane-Saulnier monoplane.
Reginald Carr began as a mechanic for Grahame-White and learnt to fly with the Grahame-White School. He went on to win many prizes. Notably the British Empire Michelin Cup when he flew a Grahame-White Type X Charabanc 315 miles on 6th November, 1914. Photo. 1914
Left — An advertisement for passenger flights from Hendon. Grahame-White can be seen at the controls of his Charabanc or "Aerobus". 1914

GRAHAME-WHITE AEROPLANES AND HYDRO-AEROPLANES

Winners of the British Michelin ... Trophy and holders of the World's Passenger Carrying Record (Pilot and Nine Passengers)

ILLUSTRATED CATALOGUES
on Application.
THE GRAHAME-WHITE AVIATION Co., Ltd.
— LONDON AERODROME, HENDON, N.W. —
or
West End Offices: 166, Piccadilly, W.

"A fine parachute descent was witnessed with R. Carr as pilot. W. Newell took his seat with his parachute on the left side of the plane. The big plane rose to 2,000 feet and when over the Edgware side of the aerodrome, Newell dropped clear. He gradually descended, landing in a field between Hendon and Mill Hill."
Hendon & Finchley Times — 5th June, 1914.

Photograph right shows a Grahame-White Charabanc or "Aerobus" with Reginald Carr seated at the controls, Louis Noel is directly behind and perched on the lower wing F. W. Gooden. Below Gooden, clutching his parachute, is the intrepid Mr. Newell.

WHITSUN HOLIDAYS AT HENDON

Parachute Experiment at Hendon—Ready to Start.

EIGHTH

LONDON AVIATION MEETING

(THREE DAYS) (Under Royal Aero Club Rules) (Weather permitting)

SATURDAY, SUNDAY and WHIT MONDAY,
MAY 30th, 31st, and JUNE 1st, 1914.

SATURDAY, MAY 30th, (3 p.m.)
Grand Speed Handicap for "The Daily Telegraph" Cup. Looping Displays and Passenger Flights.

PARACHUTE DESCENT

W. NEWELL will give the First Public Demonstration of a Parachute Descent from an Aeroplane.

SUNDAY, MAY 31st (2.30 p.m.)
Special Displays and Passenger Flights

WHIT-MONDAY, JUNE 1st (From 12 noon).
30-mile Cross Country Race for the Geisler Trophy and 100 sovs.
Looping Displays and Passenger Flights.
Twelve-Mile Speed Contest for the Hendon Cup.

"Interest in aviation shows no sign of diminution. On Bank Holiday an enormous concourse of people visited Hendon to witness flying. Should anyone have doubted the wonderful attraction to the masses exerted by the new science of aviation a visit to Hendon Aerodrome would have finally dispelled misconception."
Hendon & Finchley Times — August, 1913

During the First World War, Hendon became a major centre for aircraft manufacture, with many factories producing thousands of aeroplanes. These were built and tested here before being sent to France. The Aircraft Manufacturing Co., (see photograph below) was founded by Mr. Holt Thomas in the Edgware Road at Colindale. Its chief designer during this period (1914-18) was Mr. Geoffrey de Havilland. A little further along the Edgware Road at Cricklewood the Handley Page Co.—then in its infancy—was expanding rapidly to fulfill Government orders for a revolutionary aeroplane known as 0/100 powered by Rolls-Royce engines. This large bomber—five times the size of its contemporaries was first flown at Hendon Aerodrome on 17th December, 1915.
In addition to Airco and Handley Page, the Grahame-White Co., produced hundreds of planes including Avro 504's and D.H.6's. In 1916 the Royal Flying Corps Civilian School of Instruction was established at Hendon Aerodrome, many First World War pilots were trained here including three Victoria Cross holders, Ball, Mannock and Warneford.

Photo. below — Workers leaving the Aircraft Manufacturing Co. (Airco.) in 1917. The following aircraft were built in these factories— DH.1, DH.2, DH.3, Pro. DH.4, DH.5, DH.6, DH.9, DH.9A, DH.10, & DH.10A.

Photo. Below Right — shows a section of the Edgware Road, at Colindale. Aircraft factories can be seen on both sides of the road.

Left — The London Flying Club at Hendon, once a popular rendezvous for the wealthy. For the first few years after the First World War, Hendon led the way in the flying club movement.
This building, in later years became part of the Police College.

Below — A hansom cab race at Hendon in 1919.

Hendon Aerodrome was honoured by being chosen to stage the Victory Aerial Derby. Captain G. Gathergood flew the 190 mile course in 87 minutes 42 seconds. The winning aeroplane—seen left—was an Airco 4R, with a 450 h.p. Napier Lion engine.
Also competing in the 1919 race were Capt. J. Alcock (of Atlantic crossing fame) and H. G. Hawker, Sopwith's chief pilot.

Above — Capt. G. Gathergood (2nd left) receiving the trophy and prize after winning the 1919 Victory Aerial Derby. Grahame-White can be seen 2nd from right.

Below — A Grahame-White Bantam K.150 flown by Major R. H. Carr in the 1919 Derby. This machine had an 80 h.p. Le Rhone engine and was built in the Grahame-White factories at Hendon.

The Aerial Derby 1920 was won in record time by Mr. F. T. Courtney flying a Martinsyde Semi-Quaver at an average speed of 153½ m.p.h. The crowds at Hendon saw a dramatic finish to the race when Mr. Courtney overturned his plane on landing. His machine was badly smashed but he escaped with minor injuries.

Left — The aircraft on its back and (below) — Mr. Courtney seated on a car shortly after his miraculous escape.

In 1923 F. T. Courtney also won the First King's Cup race in a 325 h.p. Sickin Jaguar. The race began and finished at Hendon.

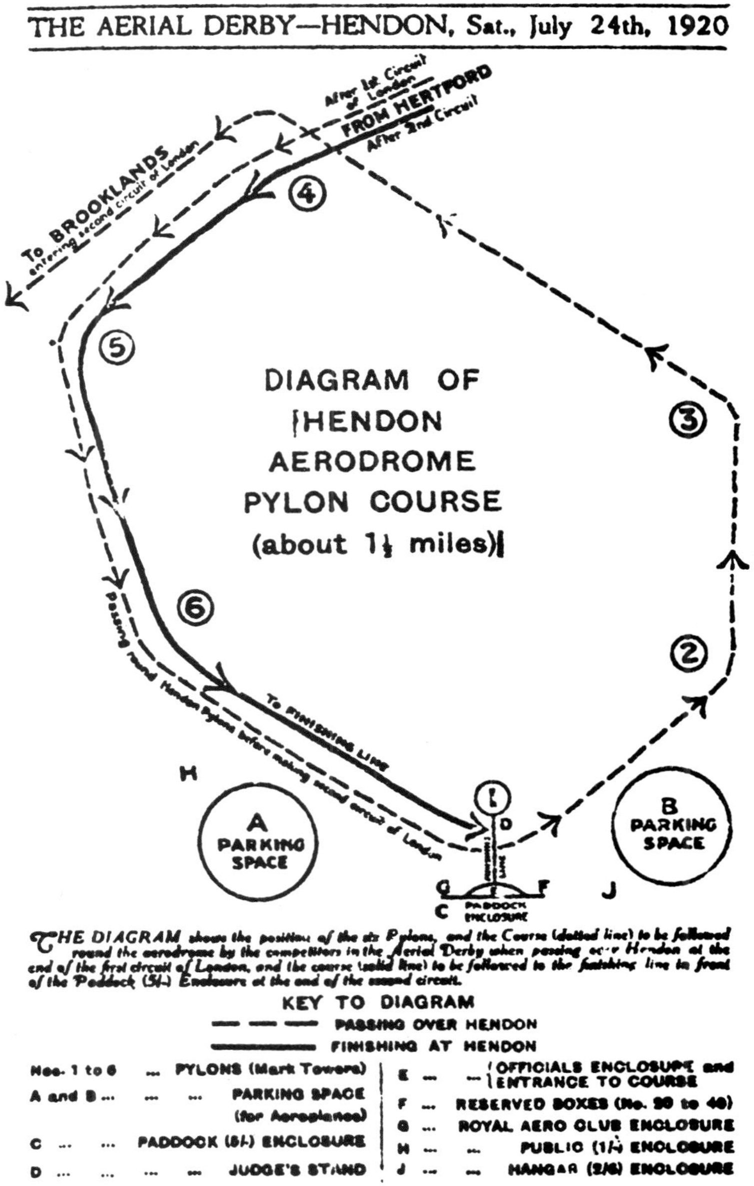

Inset above —
J. H. James taking off from Hendon to win the 1921 Aerial Derby in a Gloster Mars I (or Bamel).

Below — Hendon Air Pageant. A Bristol Fighter can be seen flying over the huge crowds gathered to see the first of the famous R.A.F. Displays. For 17 years, beginning on 5th July, 1920, the R.A.F. Tournament, known later as R.A.F. Pageant and later still as R.A.F. Display, was held at Hendon.
These R.A.F. Displays—the world's most spectacular flying shows—attracted spectators from all over the world and became the high spot of the London Season.
Some of the aeroplanes seen at the first Display were S.E.5's, Bristol Fighters, early Avros, old time Handley Page bombers, Sopwith Camels and Snipes, and Nieuports. The average speed of these aircraft was 100 m.p.h. They were the types of the last period of the Great War. Most of the pilots were men who had fought on one or another of the many Fronts.

Left — An aerial view of Hendon Aerodrome taken shortly before the Government purchased the airfield from Claude Grahame-White in 1925. The London Flying Club can be seen clearly in the foreground, with factory building on both sides of Aerodrome Road. By the end of the First World War these factories were employing three thousand men and women.

Right — Stag Lane Aerodrome, just outside the western borders of Hendon. — c. 1924
The De Havilland Aircraft Company was formed here in 1920.
Amy Johnson learnt to fly at the London Aeroplane Club, Stag Lane.
Stag Lane, like Handley Page airfield at Cricklewood and more recently Hendon, has been developed.

D.H.9A's flying in formation over Hendon Aerodrome; these photographs were taken in 1925. The picture below affords a splendid view of Hendon as it was before the construction of the Watford by-pass. Aerodrome road is clearly visible on the right of picture linking up with Hall Lane.

1927 R.A.F. PAGEANT AT HENDON
There was some spectacular flying and thrilling stunts including a huge "set piece" (an aerial destruction of a village). The R.A.F. also put on display their latest types of aircraft.
Photograph shows an Avro 504N centre of picture and on the ground to the left D.H.9A's and right Fairey Foxes (12 Squadron).

For seventeen years the R.A.F. Pageant or Display was held annually at Hendon. It was seen by more than 4 million people and raised £150,000 for R.A.F. Charities. It attracted air experts from every country in the World. Royalty attended every year and it became one of the big fashion parades of Europe.

The Displays became increasingly larger and more spectacular. Airmen leaping simultaneously from aircraft 1,000 feet up and floating safely to the ground by parachute—the sky thundering to the roar of fighting aeroplanes sweeping across the aerodrome in formation—and mock battles raging over Hendon. Dozens of planes were 'shot down' in flames, clouds of smoke would be seen coming from the back of stricken planes and the wings would burst into flames. A complicated organisation of faking was developed to achieve this effect—a stannic acid container supplied the smoke and landing flares of calcium provided the flames. There was usually a set piece—a fort or a village—which would be blown up by a man on the ground, as bombers swooped overhead.

In addition to the exhibition in the air, many new types of aircraft were first seen by the public at Hendon. In 1936 newspapers reported a new fighting aeroplane which was expected to prove the fastest in the World, designed by Mr. R. J. Mitchell.

> "The fighter known as the Spitfire will be on view for the first time at the R.A.F. Display, Hendon in June."

By 1938 however, the aircraft were becoming so fast that "show flying" near to the ground was becoming extremely difficult and dangerous, and so it was decided in the interests of safety to bring to an end the R.A.F. Displays at Hendon. It was replaced by Empire Air Day when R.A.F. stations up and down the country, including Hendon were opened to the public.

At the outbreak of World War II, fighter squadrons were based at Hendon. The fighters were moved at the commencement of the blitz in 1940. The aerodrome then became an important communications base for members of the Government and other high ranking officials. It was from Hendon that Winston Churchill made his dramatic journey to France in 1940.

After the War, flying continued at Hendon until 1957 when the airfield became too small for modern aircraft to land safely.

The R.A.F. Metropolitan Communications Squadron was transferred to Northolt on 5th November, 1957, but the service maintains a presence at Hendon which today is the home of the Joint Service Air Trooping Centre.

It was perhaps inevitable that once flying was discontinued, the airfield would be earmarked for housing and in 1968 preparations were made to begin the Grahame Park Estate which will eventually house about 10,000 people.

Hendon's flying days are over, but the crowds still wend their way down Colindale Avenue and Aerodrome Road. They flock to the R.A.F. Museum—opened by the Queen on 15th November, 1972—in their thousands, a staggering 600,000 in the first year. The museum—the only National Museum devoted entirely to aviation—occupies 10 acres of the old aerodrome and incorporates two of the original R.A.F. hangars within its structure. Among the collection of aircraft on display, ranging from a Bleriot monoplane to a Hawker Harrier, is the last aeroplane to land at Hendon. This is the Blackburn Beverley flown in on 18th June, 1968, which can now be seen outside the museum.

The Royal Air Force Museum is a splendid reminder of past glories and it would be hard to imagine a finer memorial to the history of aviation, or indeed a more fitting site than Hendon Aerodrome.

Photo — R.A.F. Display, Hendon, 1932. Crowds watching a parachute demonstration, the aircraft seen here are Vickers Virginias.